11+ Practice Papers

Multiple-Choice
Pack Three 11A

Read these instructions before you start:

- There are three sections in this paper.
- You have **50 minutes** in total to complete the whole paper. The time allowed for each section is given at the beginning of that section.
- There are **80 questions** in this paper and each question is worth **one mark**.
- You may work out the answers in rough on a separate sheet of paper.
- Answers should be marked on the answer sheet provided, not on the practice paper.
- Mark your answer in the column marked with the same number as the question by drawing a firm line clearly through the box next to your answer.
- If you make a mistake, rub it out as completely as you can and mark you new answer. You should only mark one answer for each question.
- Work as quickly and carefully as you can.
- If you find a question difficult, do **NOT** spend too much time on it but move on to the next one.
- **Calculators and protractors are not allowed.**

Moon Tuition
making the most of your potential

www.moontuition.co.uk

All rights reserved, including translation. No part of this publication may be reproduced or transmitted in any form or by any means, electronic or mechanical, including photocopying, recording, or duplication in any information storage and retrieval system, without permission in writing from the publishers, and may not be photocopied or otherwise reproduced within the terms of any licence granted by the Copyright Licensing Agency Ltd.

Mixed Multiple-Choice Pack Three 11A © Moon Tuition
www.moontuition.co.uk

Part I
Numerical Reasoning - 25 minutes

1 What is the digit 7 worth in the number 605.079?

A) 70   B) 0.7   C) 0.07   D) 7   E) 0.007

2 What is 30% of 600?

A) 18   B) 180   C) 120   D) 20   E) 200

3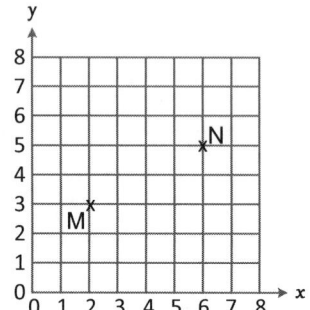

What are the coordinates of points M and N?

A) M(3,2) and N(6,5)
B) M(3,2) and N(5,6)
C) M(2,3) and N(5,6)
D) M(2,3) and N(6,5)
E) M(6,5) and N(2,3)

4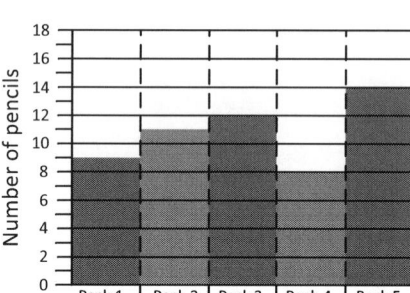

The graph shows the number of pencils in five different packs. What is the median number of pencils?

A) 14
B) 6
C) 9
D) 11
E) 8

5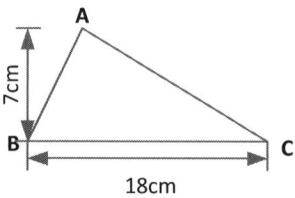
Not to scale

What is the area of the triangle ABC?

A) $126cm^2$
B) $63cm^2$
C) $25cm^2$
D) $54cm^2$
E) $136cm^2$

1

Mixed Multiple-Choice Pack Three 11A © Moon Tuition www.moontuition.co.uk

6 What is the highest common factor of 12 and 20?

A) 4   B) 60   C) 240   D) 2   E) 5

7 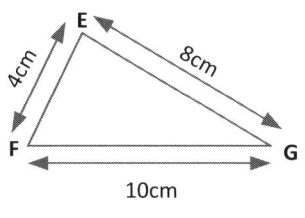 What type of triangle is this?
A) kite
B) equilateral
C) isosceles
D) rhombus
E) scalene

8 How many thirds are there in 9?

A) 3   B) 4   C) 27   D) 2   E) $\frac{1}{3}$

9 What is $\frac{2}{3}$ of 120?

A) 40   B) 80   C) 60   D) 120   E) 30

10 There are 36 students in a class. Out of all the students, $\frac{1}{4}$ like Latin, $\frac{1}{2}$ like English, the rest like Greek. How many students like Greek?

A) 9   B) 18   C) 27   D) 12   E) 24

11 What is the lowest common multiple of 8 and 12?

A) 4   B) 48   C) 96   D) 24   E) 12

12 There are 3 green, 5 blue and 4 red marbles in a bag. What is the probability of picking a green marble from the bag?

A) $\frac{3}{10}$ B) 35%   C) $\frac{1}{4}$ D) $\frac{1}{3}$ E) 0.2

13 What is $9000 + 90 + 9 + 0.9$ in standard format?

A) 9099.9   B) 9990.9   C) 9999.9   D) 9099.09   E) 9909.9

14 There are twenty-eight students in Jack's class. Twenty-one of them are girls. What is the proportion of boys in Jack's class?

A) $\frac{3}{4}$ B) $\frac{21}{28}$ C) $\frac{1}{4}$ D) $\frac{1}{3}$ E) 75%

15 Round 24509 to the nearest thousand.

A) 24000 B) 24500 C) 30000 D) 24600 E) 25000

16 Work out $\frac{7}{10} - \frac{3}{10}$
Express your answer in its simplest form.

A) $\frac{4}{10}$ B) $\frac{2}{5}$ C) $\frac{10}{10}$ D) $\frac{1}{10}$ E) $\frac{2}{10}$

17 There are 70 marbles in a box. 10% of them are green, 30% are blue, the rest are red. How many red marbles are there in the box?

A) 60 B) 40 C) 28 D) 42 E) 50

18 A tissue box measures 30cm long, 12cm wide and 10cm high. What is the volume of the box?

A) $3600cm^3$ B) $360cm^3$ C) $120cm^3$ D) $12000cm^3$ E) $36cm^3$

19 The following table shows the highest temperature recorded in Miami from April to September. What is the mode?

April	May	June	July	August	September
28 °C	30 °C	32 °C	35 °C	36 °C	30 °C

A) 8 °C B) 30 °C C) 36 °C D) 32 °C E) 28 °C

20 You are told $245 \times 36 = 8820$. What is 245×18?

A) 17640 B) 4440 C) 16640 D) 4420 E) 4410

Mixed Multiple-Choice Pack Three 11A

© Moon Tuition
www.moontuition.co.uk

21 This machine takes a number as input, multiplies it by 25% and then takes 4 away. Which number comes out?

A) 10   B) 6   C) 14   D) 4   E) 8

22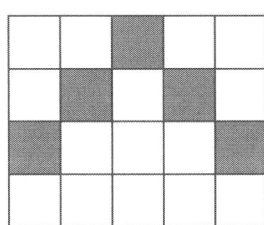

What percentage of the diagram is unshaded?

A) 25%
B) 30%
C) 75%
D) 60%
E) 80%

23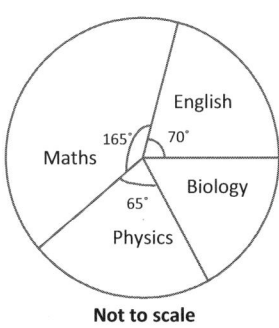

A group of 60 students were asked what their favourite subjects were. How many students said Biology?

A) 60
B) $\frac{1}{6}$
C) 10
D) 15
E) 40

24 What percentage of 1 hour is 15 minutes?

A) $\frac{1}{4}$ B) 20%   C) $\frac{2}{3}$ D) 30%   E) 25%

25 A school spent £24 to buy pens. Each pen costs 80p. How many pens did the school buy?

A) 3   B) 4   C) 20   D) 30   E) 6

26 What is 4.5×5.6?

A) 252   B) 25.2   C) 2.52   D) 0.252   E) 26.2

Mixed Multiple-Choice Pack Three 11A

27 What is 400 x 800?

A) 320000   B) 32000   C) 3200   D) 3200000   E) 32000000

28

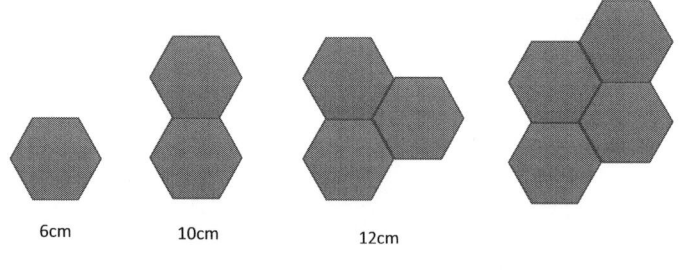

6cm   10cm   12cm
Not to scale

Here are 4 shapes made from regular hexagons. If the perimeters of the first three shapes are: 6cm, 10cm and 12cm. What is the perimeter of the fourth shape?

A) 24cm   B) 20cm   C) 16cm   D) 15cm   E) 14cm

29 Mark is throwing a 1-6 dice. Which one of the following statements is true?

A) The probability of getting an odd number is $\frac{2}{3}$.
B) The probability of getting a prime number is 40%.
C) The probability of getting an even number is 50%.
D) The probability of getting 4 is $\frac{1}{4}$.
E) The probability of getting 6 number is $\frac{1}{5}$.

30 You multiply a number by itself. The answer is then multiplied by the number you started with. The new number is 27. What number did you start with?

A) 9   B) 3   C) 27   D) 1   E) 5

THE END OF NUMERICAL REASONING PART

Part II
Verbal Reasoning - 15 minutes

Section 1 - Comprehension Test

Read the poem carefully and then answer the questions that follow.

Dulce et Decorum Est

Bent double, like old beggars under sacks,
Knock-kneed, coughing like hags, we cursed through sludge,
Till on the haunting flares we turned our backs
And towards our distant rest began to trudge.
Men marched asleep. Many had lost their boots
But limped on, blood-shod. All went lame; all blind;
Drunk with fatigue; deaf even to the hoots
Of tired, outstripped Five-Nines1
that dropped behind.
Gas! Gas! Quick, boys! An ecstasy of fumbling,
Fitting the clumsy helmets just in time;
But someone still was yelling out and stumbling
And flound'ring like a man in fire or lime...
Dim, through the misty panes and thick green light,
As under a green sea, I saw him drowning.
In all my dreams, before my helpless sight,
He plunges at me, guttering, choking, drowning.
If in some smothering dreams you too could pace
Behind the wagon that we flung him in,
And watch the white eyes writhing in his face,
His hanging face, like a devils sick of sin;
If you could hear, at every jolt, the blood
Come gargling from the froth-corrupted lungs,
Obscene as cancer, bitter as the cud
Of vile, incurable sores on innocent tongues,
My friend, you would not tell with such high zest
To children ardent for some desperate glory,
The old Lie: Dulce et decorum est
Pro patria mori1

By Wilfred Owen, 1893 - 1918

Mixed Multiple-Choice Pack Three 11A © Moon Tuition
www.moontuition.co.uk

31 This poem is describing which of the following:

A) An intense rugby match
B) A soldier's experience of World War I
C) A boxing match
D) Hunting deer
E) A nurse's experience of World War I

32 'Like old beggars' (line 1) is an example of which of the following?

A) A metaphor
B) A noun
C) A simile
D) An onomatopoeia
E) A verb

33 'Men marched asleep' (line 5) shows that:

A) The men find it hard to stay awake while marching due to fatigue
B) The men are literally marching with their eyes closed
C) The men gained energy to march from sleeping
D) The men liked to sleepwalk
E) The men are frustrated

34 'Deaf even to the hoots of disappointed shells' means that:

A) The men are so tired they cannot even hear the sound of gunfire
B) The men cannot hear because of sea shells they step on
C) The sea creatures with shells are disappointed when the men march by
D) The sea creatures with shells are deaf
E) The men are deafened from gunfire

35 'Gas' (line 10) is most likely which of the following:

A) The air the men breathe in
B) Oxygen received through their helmets
C) The toxic gas bombs the enemy have thrown at the narrator
D) The toxic gas bombs the narrator has thrown at the enemy
E) The gas used in a cooking stove

Mixed Multiple-Choice Pack Three 11A © Moon Tuition
www.moontuition.co.uk

36 The word 'Fitting' in line 11 is an example of:

A) A proper noun
B) A participle
C) A simile
D) An abstract noun
E) A common noun

37 'A green sea' (line 14) is a metaphor for which of the following:

A) The green sea the men can see in the distance
B) The vomit from the men caused by the toxic gas
C) The men are all wearing green so that it looks like a sea of green
D) The green toxic gas is so thick that it looks like a green sea
E) None of the above

38 'Him' in line 14 is an example of:

A) An adjective
B) A participle
C) An adverb
D) A pronoun
E) None of the above

39 Why is the man 'choking' in line 16:

A) The man has inhaled the toxic gas
B) The man choked on some food
C) The man swallowed some shrapnel from a bomb
D) The man swallowed water from the sea
E) The man has asthma

40 'Obscene' in line 23 means:

A) Healthy
B) strange
C) Disgusting
D) Inefficient
E) Happy

Mixed Multiple-Choice Pack Three 11A

© Moon Tuition
www.moontuition.co.uk

Section 2 - Synonyms

Select the word which has the SIMILAR meaning as the word in bold. Mark your answer on the answer sheet by choosing one of the options A-E.

41
desire

    A. overtake   B. long for   C. desert   D. sever   E. massive

42
isolated

    A. land   B. meaning   C. secluded   D. faith   E. fundamental

43
dear

    A. forceful   B. ally   C. friend   D. expensive   E. hate

44
feeble

    A. weak   B. strong   C. tent   D. flexible   E. edible

45
singe

    A. route   B. confuse   C. detect   D. burn   E. consent

PLEASE CONTINUE TO THE NEXT PAGE

46
hoax

    A. amble   B. trick   C. enable   D. elated   E. failure

47
valiant

    A. courageous   B. variable   C. solute   D. safe   E. empty

48
bashful

    A. grateful   B. deceive   C. shy   D. excited   E. strive

49
squander

    A. realise   B. courage   C. utilize   D. waste   E. modify

50
surrender

    A. widen   B. guard   C. capitulate   D. vision   E. respond

Section 3 - Shuffled Sentence

The following are shuffled sentences. Select the word that needs to be removed so that you can rearrange the rest of the words to make a sentence. Mark your answer on the answer sheet by choosing one of the options A-E.

51 enjoys the in pool very Henry swimming

A) the   B) in   C) very   D) Henry   E) enjoys

52 saw bats, they elephants, and monkeys a

A) they   B) bats   C) and   D) elephants   E) a

53 the garden feeds the birds Jack in space

A) space   B) the   C) birds   D) in   E) Jack

54 mum money happens took her bag of out some

A) money   B) happens   C) took   D) some   E) bag

55 dog injection vet the gave an roller the

A) dog   B) vet   C) the   D) roller   E) gave

56 will what be Jasmine wearing in tonight

A) in   B) tonight   C) Jasmine   D) wearing   E) be

THE END OF VERBAL REASONING PART

Mixed Multiple-Choice Pack Three 11A © Moon Tuition
www.moontuition.co.uk

Part III
Non Verbal Reasoning - 10 minutes

Section 1

There are two shapes on the left of each of the rows below. The second is related to the first in some way. There are five shapes on the right. Find which one of the five shapes is related to the third shape in the same way as the two on the left. Mark its letter on the answer sheet.

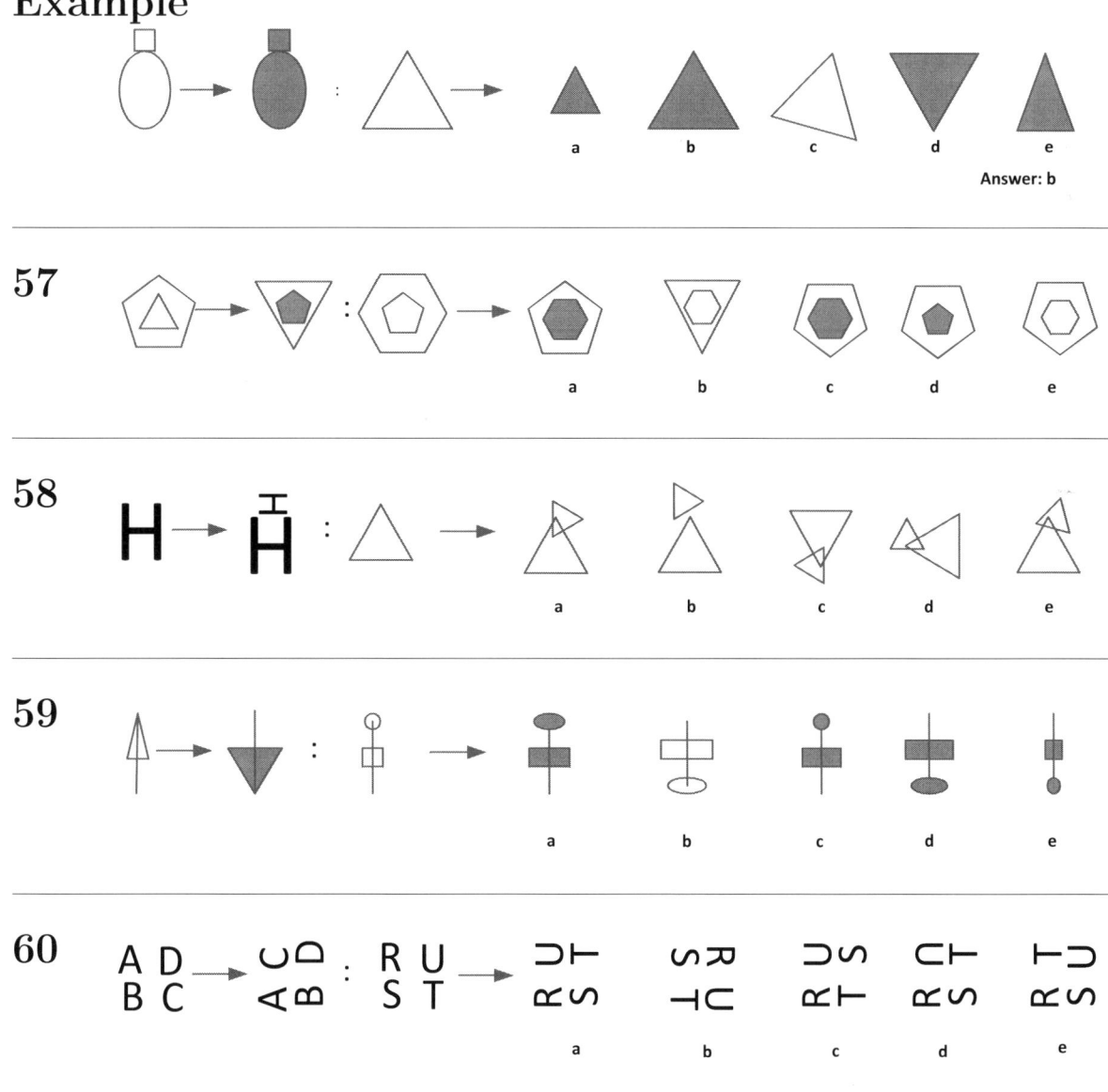

Mixed Multiple-Choice Pack Three 11A

61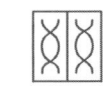
                                                                                     a   b   c   d   e

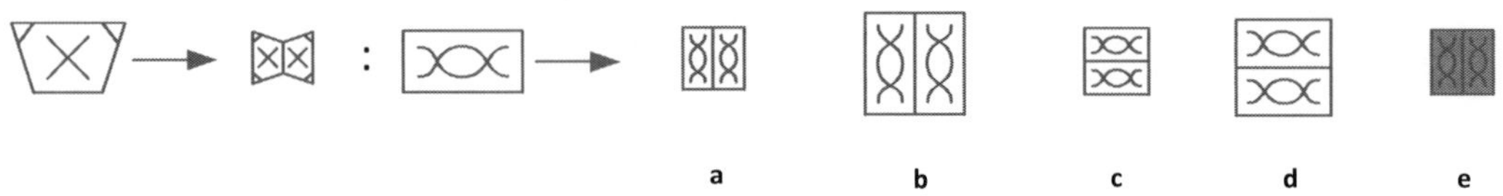

PLEASE CONTINUE TO THE NEXT PAGE

62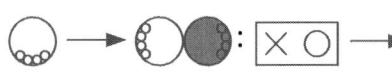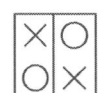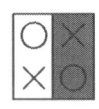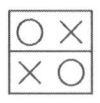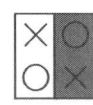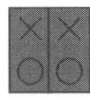
                                                                                     a   b   c   d   e

63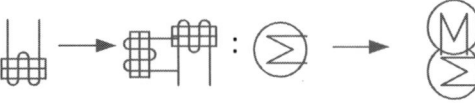
                                                                                     a   b   c   d   e

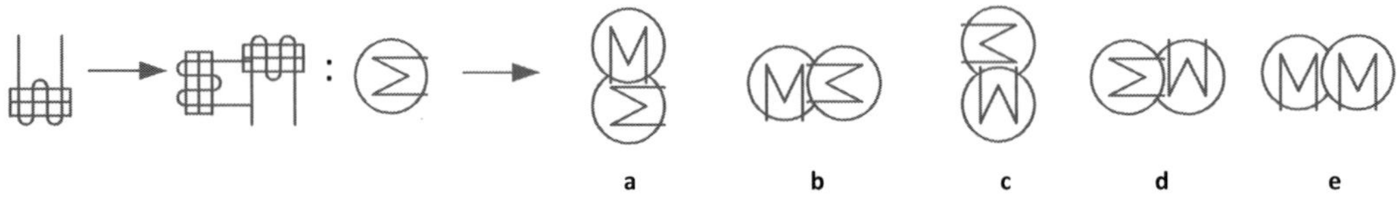

64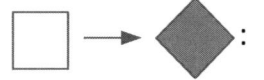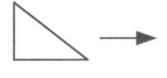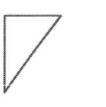
                                                                                     a   b   c   d   e

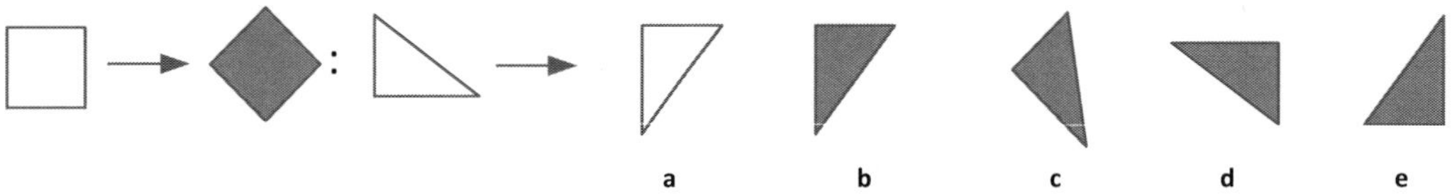

65
                                                                                     a   b   c   d   e

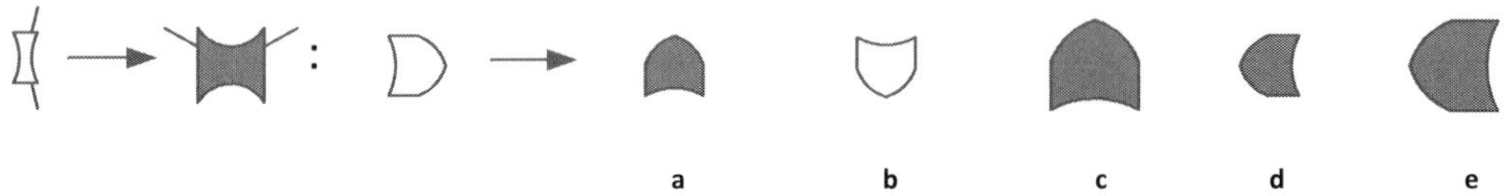

66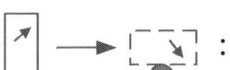
                                                                                     a   b   c   d   e

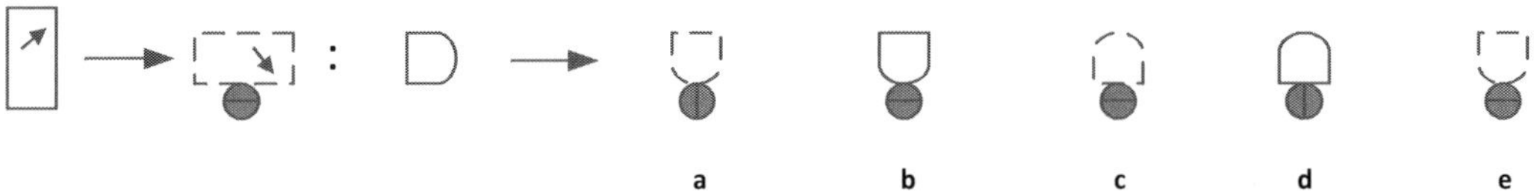

67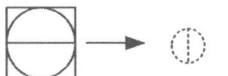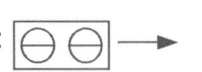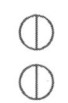
                                                                                     a   b   c   d   e

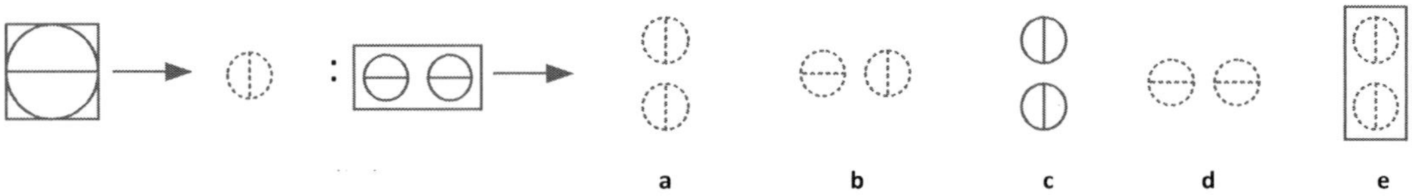

Mixed Multiple-Choice Pack Three 11A

PLEASE CONTINUE TO THE NEXT PAGE

68

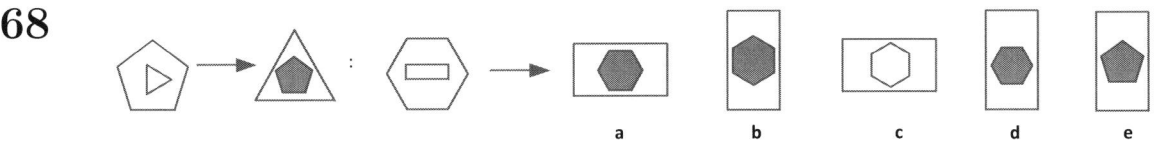

PLEASE CONTINUE TO THE NEXT PAGE

Mixed Multiple-Choice Pack Three 11A © Moon Tuition www.moontuition.co.uk

Section 2

There are smaller squares in each large square on the left. One of the small squares has been left empty. Work out which one of the five figures on the right can fill the empty square and mark its letter on the answer sheet.

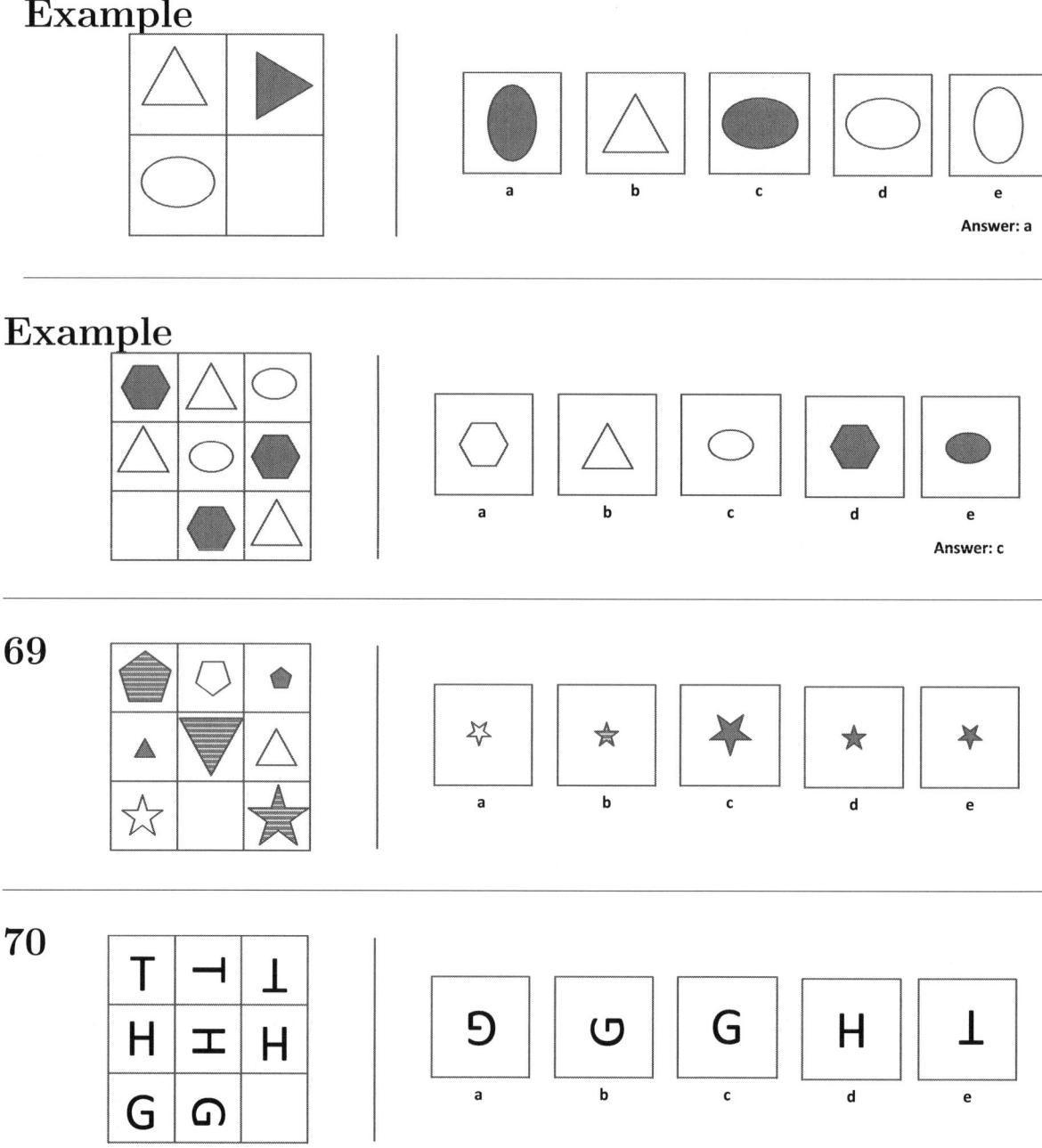

Mixed Multiple-Choice Pack Three 11A © Moon Tuition www.moontuition.co.uk

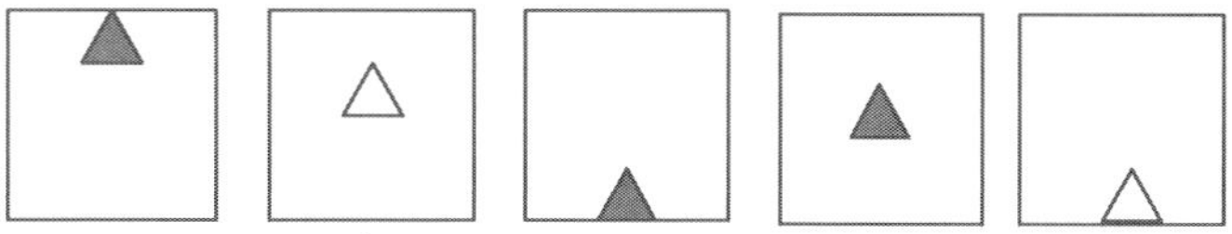

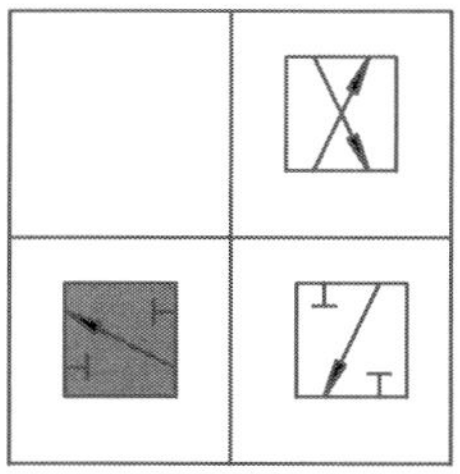

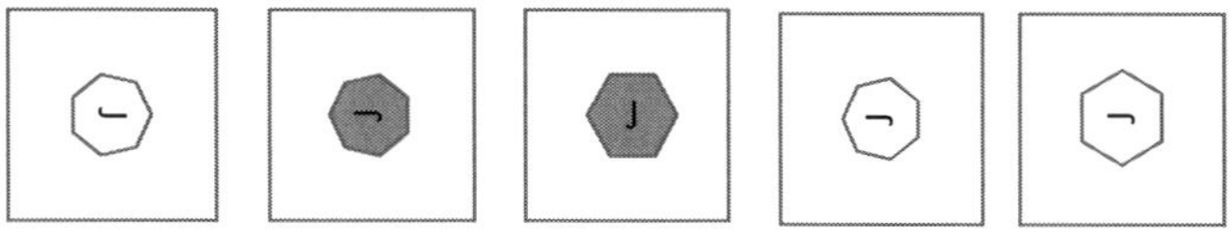

71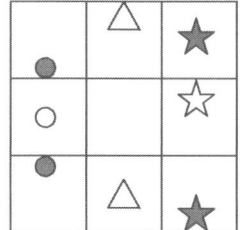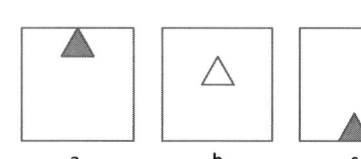

72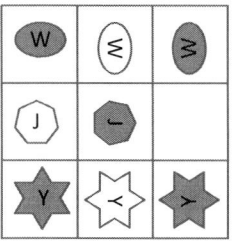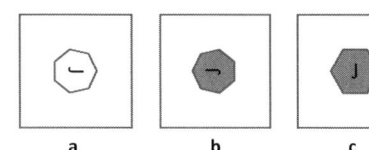

73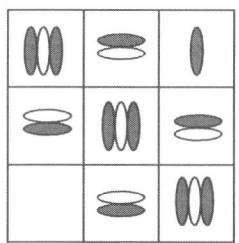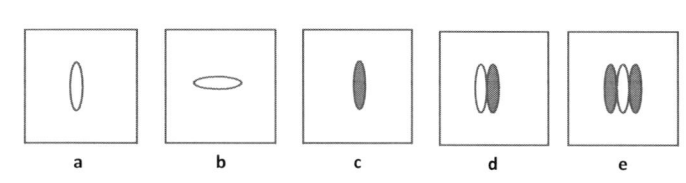

74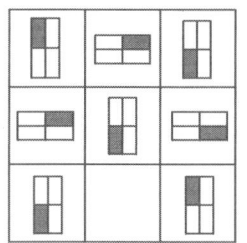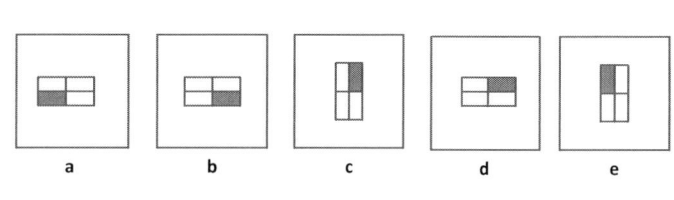

75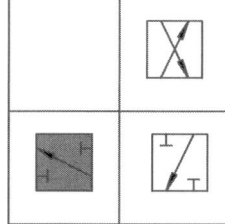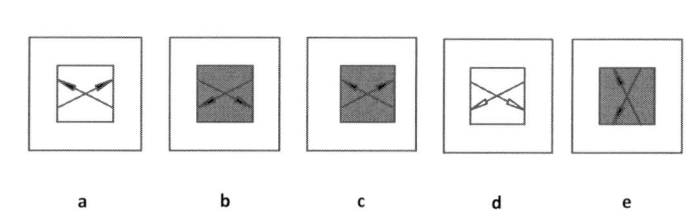

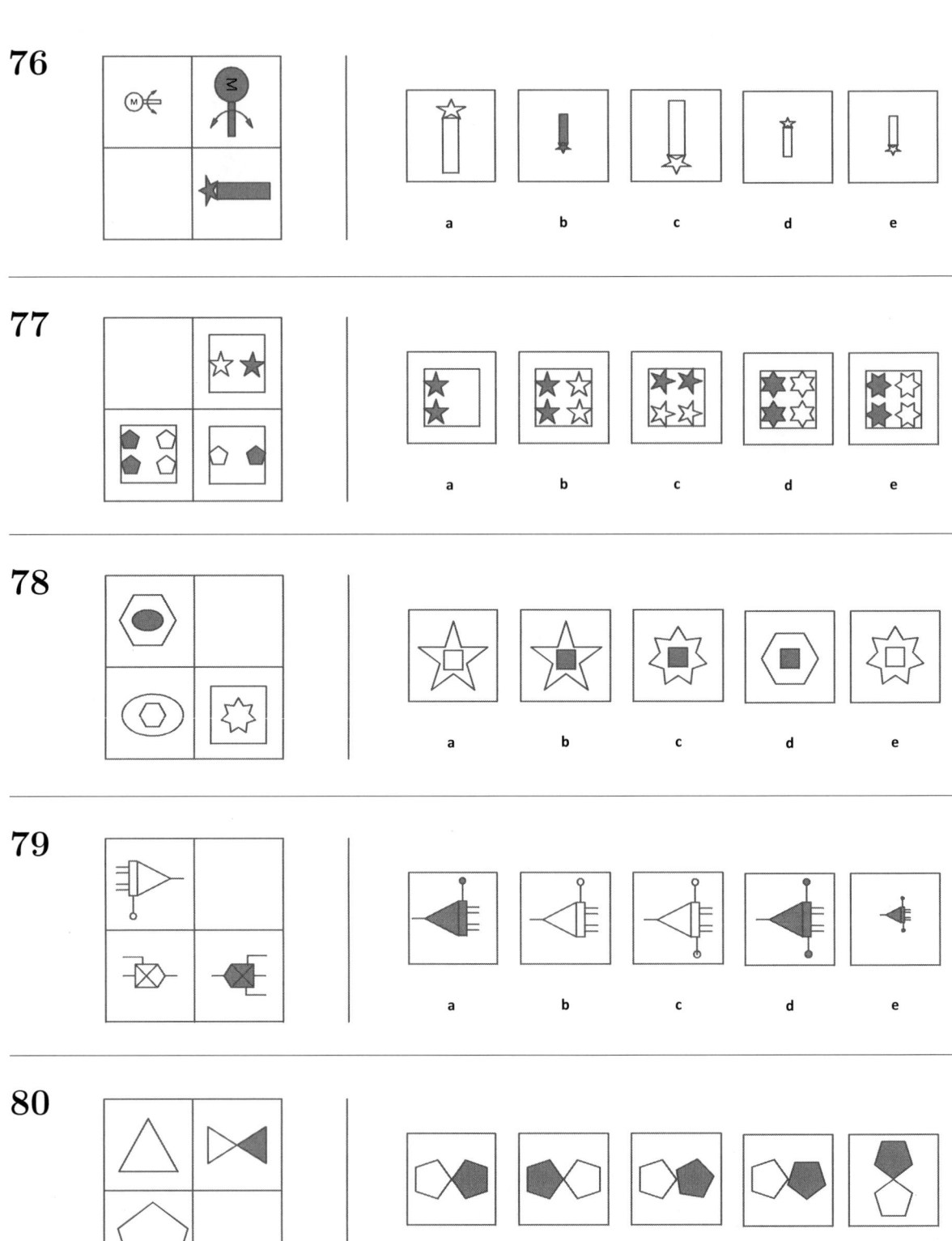

THE END OF NON VERBAL REASONING PART

THE END OF TEST PAPER 11A

11+ Practice Papers

Multiple-Choice
Pack Three 11B

Read these instructions before you start:

- There are three sections in this paper.
- You have **50 minutes** in total to complete the whole paper. The time allowed for each section is given at the beginning of that section.
- There are **80 questions** in this paper and each question is worth **one mark**.
- You may work out the answers in rough on a separate sheet of paper.
- Answers should be marked on the answer sheet provided, not on the practice paper.
- Mark your answer in the column marked with the same number as the question by drawing a firm line clearly through the box next to your answer.
- If you make a mistake, rub it out as completely as you can and mark you new answer. You should only mark one answer for each question.
- Work as quickly and carefully as you can.
- If you find a question difficult, do **NOT** spend too much time on it but move on to the next one.
- **Calculators and protractors are not allowed.**

Moon Tuition
making the most of your potential
www.moontuition.co.uk

All rights reserved, including translation. No part of this publication may be reproduced or transmitted in any form or by any means, electronic or mechanical, including photocopying, recording, or duplication in any information storage and retrieval system, without permission in writing from the publishers, and may not be photocopied or otherwise reproduced within the terms of any licence granted by the Copyright Licensing Agency Ltd.

Mixed Multiple-Choice Pack Three 11B © Moon Tuition
www.moontuition.co.uk

Part IV
Numerical Reasoning - 25 minutes

1 What is the next number in the sequence below?

10   25   40   55   70

A) 80   B) 95   C) 75   D) 85   E) 90

2 Benedict has £3.60. How many 40p chocolates can he buy?

A) 8   B) 10   C) 9   D) 90   E) 7

3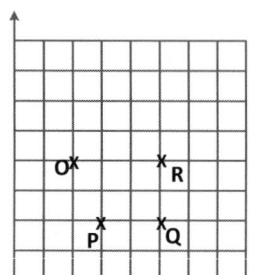

What is the shape when points O, P, Q and R are joined in order?

A) Parallelogram
B) Kite
C) Rectangle
D) Rhombus
E) Trapezium

4 Granny's weighing scales are not working properly. When you weigh something it says 4 grams less than it should be. If 4 oranges weigh 28 grams on the scales, how many grams does each orange actually weigh?

A) 32   B) 24   C) 6   D) 12   E) 8

5 Round 34.8557 to the nearest hundredth.

A) 34.85   B) 34.86   C) 34.9   D) 34.855   E) 34.8

6 What is $\frac{1}{3}$ of four hundred and two?

A) 134   B) 144   C) 135   D) 1206   E) 133

1

7 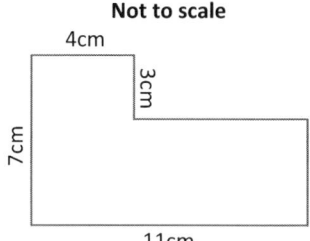 What is the area of this shape?

A) $36cm^2$
B) $56cm^2$
C) $18cm^2$
D) $53cm^2$
E) $44cm^2$

8 A laptop costs £400. There will be a 20% off sale this weekend, how much will the sale price be this weekend?

A) £80   B) £480   C) £320   D) £392   E) £240

9 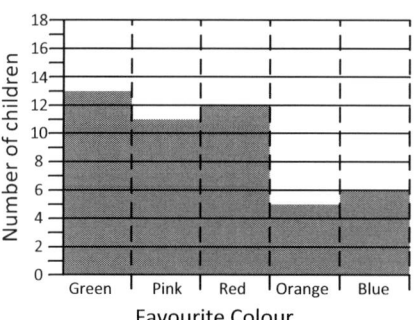 The graph shows the favourite colour of a group of children. How many more children like green than blue?

A) 7
B) 6
C) 8
D) 5
E) 9

10 What is 8^2?

A) 16   B) 4   C) 54   D) 64   E) 36

11 Jason is $1.63m$ tall. What is his height in centimetres?

A) 16.3   B) 163   C) 1630   D) 1.63   E) 0.163

12 Which one of the following is the biggest?

A) 68%   B) $\frac{3}{5}$ C) $\frac{2}{3}$ D) $\frac{3}{4}$ E) 0.81

Mixed Multiple-Choice Pack Three 11B

13 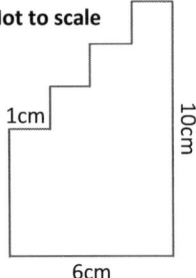 What is the perimeter of this shape?

A) 16cm
B) 32cm
C) 24cm
D) 20cm
E) 18cm

14 Which of these fractions is equivalent to $\frac{4}{9}$?

A) $\frac{2}{3}$ B) $\frac{8}{16}$ C) $\frac{5}{9}$ D) $\frac{8}{18}$ E) $\frac{12}{36}$

15 Round 4569 to the nearest 100.

A) 4560   B) 4570   C) 4600   D) 4500   E) 5000

16 In a chess club, there are 3 boys for every 2 girls. There are 15 children altogether. How many girls are there?

A) 6   B) 9   C) 7   D) 5   E) 10

17 What is the missing number in this table?

IN	2	4	5	7	8
OUT	6	12	15	21	

A) 16   B) 26   C) 24   D) 32   E) 40

18 A box can hold 20 oranges. How many boxes are needed for 168 oranges?

A) 8   B) 9   C) 7   D) 10   E) 11

19 Tom wants to buy 3 chocolate bars. Each chocolate bar costs 40p. How much change from a £5 note will he get?

A) £1.20   B) £4.80   C) £3.80   D) £2.80   E) £3.60

20 I think of a number, multiply it by 7 and then add 4. The answer is 46. What was the number I thought of?

A) 8 B) 4 C) 5 D) 7 E) 6

21 Emily has 2400g of flour. She needs $\frac{3}{4}$ of it to make a cake. How many grams of flour does she need?

A) 600 B) 60 C) 180 D) 1800 E) 1600

22 What is the smallest number you can make from these four digits?

5 9 6 1

A) 5961 B) 9561 C) 1596 D) 1659 E) 1569

23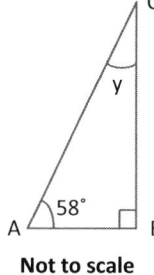

What is the size of angle **y**?

A) 32°
B) 42°
C) 138°
D) 148°
E) 34°

Not to scale

24 A map of England is drawn to a scale of 1:800,000. The distance between Harrow and Reading is $64km$. How many centimetres on the map would represent this distance?

A) 2cm B) 1.6cm C) 8cm D) 80cm E) 0.8cm

25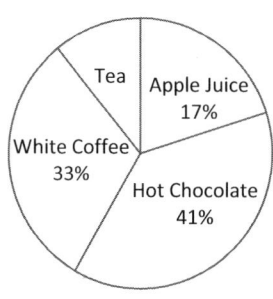

The pie chart shows the percentage of drinks sold in a shop last Saturday. What percentage represents the tea sold?

A) 91%
B) 8%
C) 9%
D) 18%
E) 7%

Mixed Multiple-Choice Pack Three 11B　　　© Moon Tuition　www.moontuition.co.uk

26　What is the range of the following data list?

0.13　　3.78　　6.43　　0.09　　11.05　　5.9　　3.78　　4.66

A) 10.96　　B) 10.92　　C) 3.78　　D) 4.66　　E) 11.05

27　Jim has fencing training twice a week. The cost is £11.50 per session. What is the total cost of his fencing training for 8 weeks?

A) £23　　B) £184　　C) £92　　D) £194　　E) £174

28　How many prime numbers are there between 20 and 30?

A) 0　　B) 4　　C) 1　　D) 3　　E) 2

29　What do we call an angle that is between 90° and 180°?

A) Right Angle　B) Isosceles　C) Acute Angle　D) Reflex Angle　E) Obtuse Angle

30　At 1 am, the temperature was -9°C. At 11 am, it was 2°C. What was the difference in temperature?

A) -11°C　　B) 7°C　　C) 11°C　　D) -7°C　　E) 10°C

THE END OF NUMERICAL REASONING PART

Part V
Verbal Reasoning - 15 minutes

Section 1 - Comprehension Test

Read the poem carefully and then answer the questions that follow.

The following poem describes the events of a painful farewell between a man and his lover (the reason of the separation is unexplained). The two are currently standing on a platform at a train station the man watches as his lover boards the train and gradually disappears from view.

On the Departure Platform
by Thomas Hardy

We kissed at the barrier; and passing through
She left me, and moment by moment got
Smaller and smaller, until to my view
She was but a spot;

A wee white spot of muslin(1) fluff
That down the diminishing platform bore
Through hustling crowds of gentle and rough
To the carriage door.

Under the lamplight's fitful glowers,
Behind dark groups from far and near,
Whose interests were apart from ours,
She would disappear,

Then show again, till I ceased to see
That flexible form, that nebulous white;
And she who was more than my life to me
Had vanished quite.

We have penned(2) new plans since that fair fond day,
And in season she will appear again
Perhaps in the same soft white array
But never as then!

Mixed Multiple-Choice Pack Three 11B

© Moon Tuition
www.moontuition.co.uk

'And why, young man, must eternally fly(3)
A joy you'll repeat, if you love her well?'
O friend, nought(4) happens twice thus(5); why,
I cannot tell!

1. 'muslin' n. a lightweight cotton material
2. 'to pen' v. to write down/ to create/ to formulate (in this context)
3. 'to fly' v. to vanish/ disappear/ pass by/ go away/ (in this context)
4. 'nought' pron. nothing
5. 'thus' adv. in this way/ like this/ in such a way

31 Which of the following occurs in the first stanza?

A) The woman walks through the man
B) The man and the woman both get on the train and kiss
C) The man starts crying as the woman boards the train
D) The woman gradually diminishes in size as she moves away on the train
E) The woman starts to grow spots

32 The meaning of the word 'wee' in this context (line 1, stanza 2) is:

A) Large
B) Tiny
C) Invisible
D) Grey
E) Milky

33 The meaning of the word 'bore' (past tense of 'bore') in this context (line 2, stanza 2) is:

A) to move swiftly and inexorably through something
B) to become bored
C) a wild pig
D) to move slowly and gradually forward without purpose or momentum
E) to drag oneself through something reluctantly

34 The actions of the members of the crowd in line 3 of stanza 2 are best described as:

A) running past one another
B) hitting and punching one another
C) pushing and jostling past one another
D) climbing over one another
E) hurrying past one another desperately

35 What do you think the word 'nebulous' means in line 2 of stanza 4?

A) Dark
B) Solid
C) Tangible
D) Invisible
E) Hazy

36 Which of the following best reveals the man's strong love for the woman?

A) We kissed at the barrier
B) A wee white spot of muslin fluff
C) That flexible form, that nebulous white
D) And she who was more than my life to me
E) And in season she will appear again

37 The phrase 'in season' (line 2, stanza 5) means:

A) In Spring
B) In Autumn
C) In Winter
D) Hopefully
E) In the future

38 The word 'array' (line 3, stanza 5) in this context means:

A) Clothing
B) Appearance
C) Display
D) An ordered arrangement
E) Exhibition

Mixed Multiple-Choice Pack Three 11B © Moon Tuition
www.moontuition.co.uk

39 Which of the following do you think best explains the meaning of the last stanza?

A) The man will always be flying
B) Joyous moments will never last, nothing good ever happens more than once.
C) Joyous moments will always last, good things will happen all the time.
D) Happy moments are flying in the air
E) Love is flying in the air

40 The word 'passing' in stanza 1 is a:

A) Participle
B) Preposition
C) Adverb
D) Adjective
E) Proper noun

41 What part of speech is 'Under' in stanza 3?

A) Proper noun
B) Abstract noun
C) Common noun
D) Preposition
E) Adjective

42 The word 'white' in stanza 2 is an example of:

A) A noun
B) A participle
C) A relative pronoun
D) An adjective
E) The definite article

43 The word 'ours' is a:

A) Possessive pronoun
B) Adjective
C) Noun
D) Adverb
E) Preposition

9

44 The poem is entitled 'On the Departure Platform' however trains can arrive as well as depart from the same platform. The word 'Departure' is therefore unnecessary but is used anyway. Which of the following is most likely the reason?

A) It emphasises the man's relief that his lover is finally leaving
B) It emphasises the man's sense of loss as his lover boards the train
C) It creates a special rhythm in the title of the poem
D) It shows that the man's lover is happy to leave
E) It shows that the trains are moving

Mixed Multiple-Choice Pack Three 11B © Moon Tuition
www.moontuition.co.uk

Section 2 - Missing word

In each of the following sentences, there is a word missing. Please complete each sentence by selecting one word from the options A-E. Mark your answer on the answer sheet.

45 Jessica enjoys( ) in her garden.

A) walk B) walking C) to playing D) to have been E) play

46 This includes ( ) the limits on places for applicants who achieve AAB grades or better.

A) remove B) removes C) removing D) to remove E) removeing

47 ( ) also been concerns about the instability.

A) There have B) Their C) Some D) They E) Their have

48 The number ( ) for the first time in recent years.

A) haven't risen B) hasn't risen C) have risen D) has rose E) have been risen

49 This is one of ( ) great growth industries of the future.

A) britain's B) Britain's C) britain D) england E) london

50 Parents are( ) they have fewer than 5 hours a week to play with their children.

A) so busy that B) such busy C) more busy D) been busy E) such busy which

PLEASE CONTINUE TO THE NEXT PAGE

Section 3 - Shuffled Sentence

The following are shuffled sentences. Select the word that needs to be removed so that you can rearrange the rest of the words to make a sentence. Mark your answer on the answer sheet by choosing one of the options A-E.

51 around only third a of children I played with objects household

A) children   B) I   C) played   D) with   E) objects

52 should the government safe to develop outdoor spaces

A) develop   B) spaces   C) to   D) should   E) outdoor

53 more should children in play outside

A) in   B) more   C) play   D) outside   E) children

54 how that's feel I on music about

A) how   B) that's   C) music   D) on   E) about

55 won't me bother it with start to all

A) won't   B) bother   C) all   D) it   E) with

56 certainly you be will has challenged

A) than   B) no   C) kitchen   D) has   E) by

THE END OF VERBAL REASONING PART

Mixed Multiple-Choice Pack Three 11B

© Moon Tuition
www.moontuition.co.uk

Part VI
Non Verbal Reasoning - 10 minutes

Section 1

Which one of the five shapes completes the sequence? Mark its letter on the answer sheet.

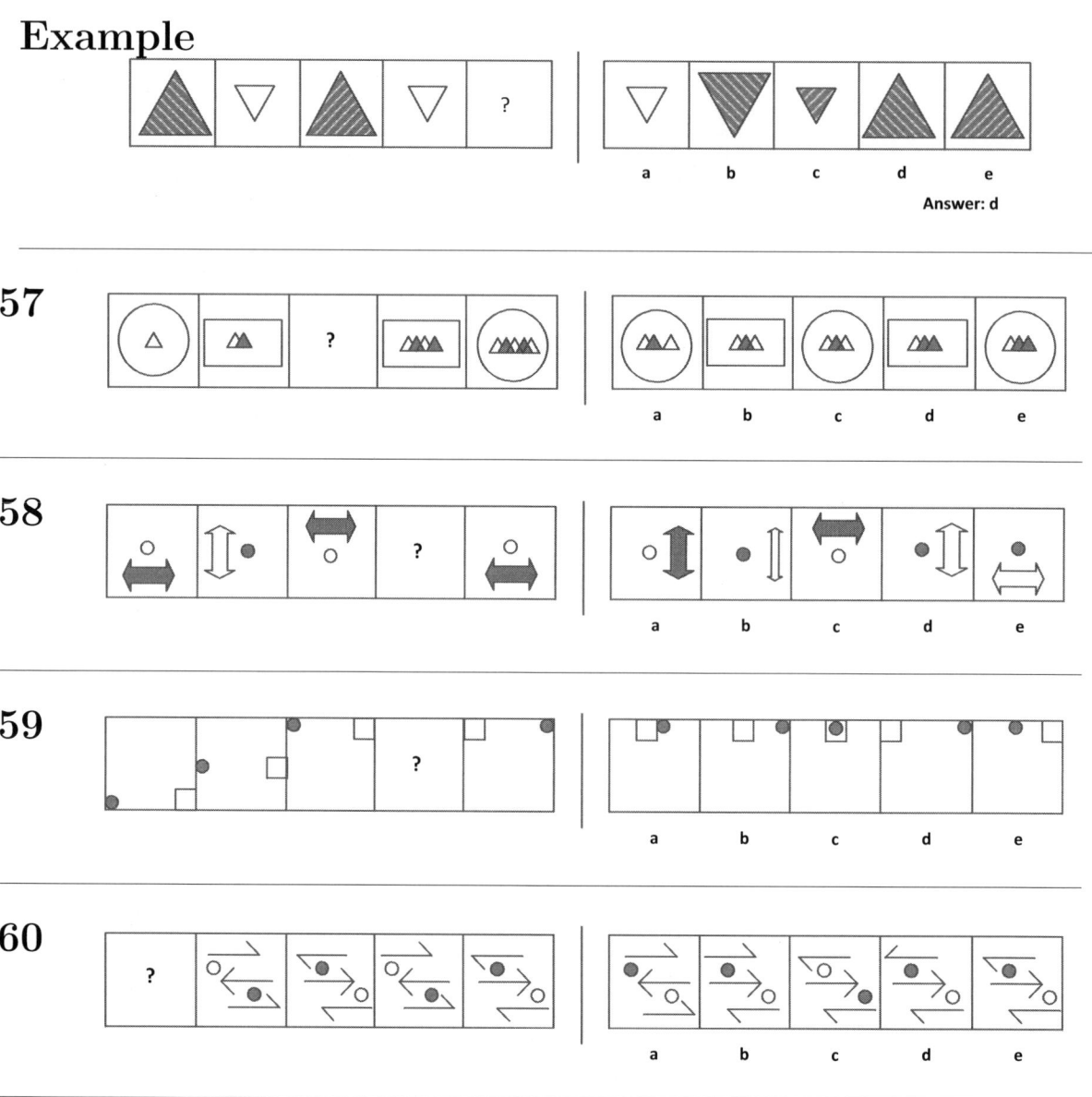

PLEASE CONTINUE TO THE NEXT PAGE

Mixed Multiple-Choice Pack Three 11B

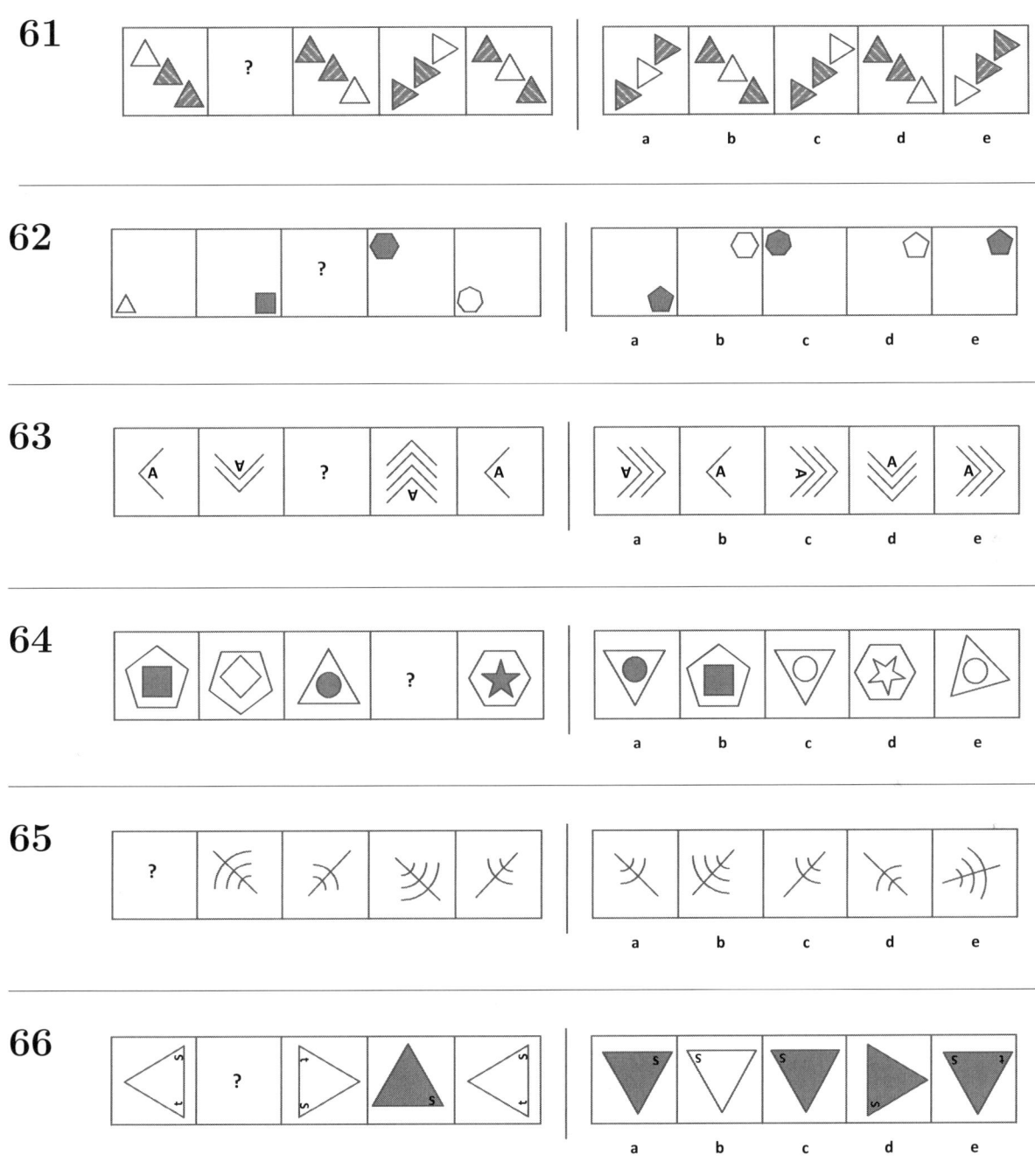

Mixed Multiple-Choice Pack Three 11B

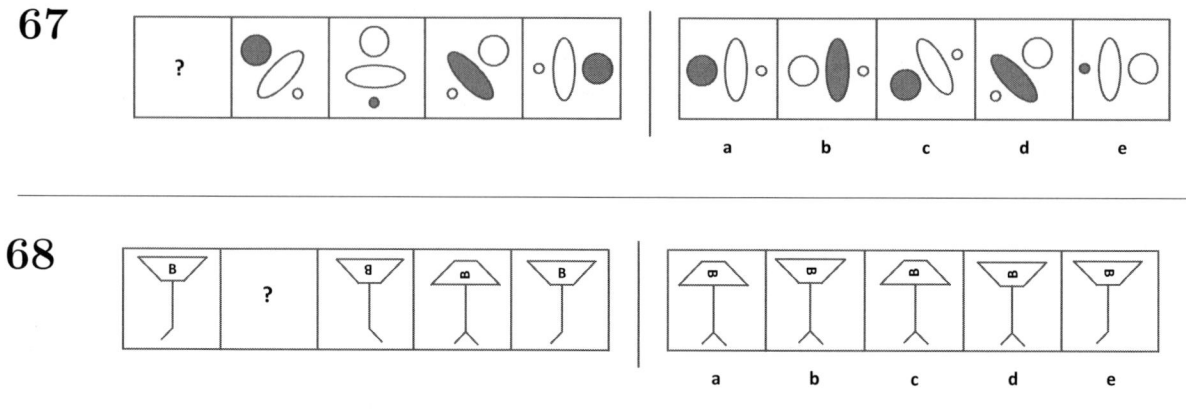

PLEASE CONTINUE TO THE NEXT PAGE

Mixed Multiple-Choice Pack Three 11B

© Moon Tuition
www.moontuition.co.uk

Section 2

There are smaller squares in each large square on the left. One of the small squares has been left empty. Work out which one of the five figures on the right can fill the empty square and mark its letter on the answer sheet.

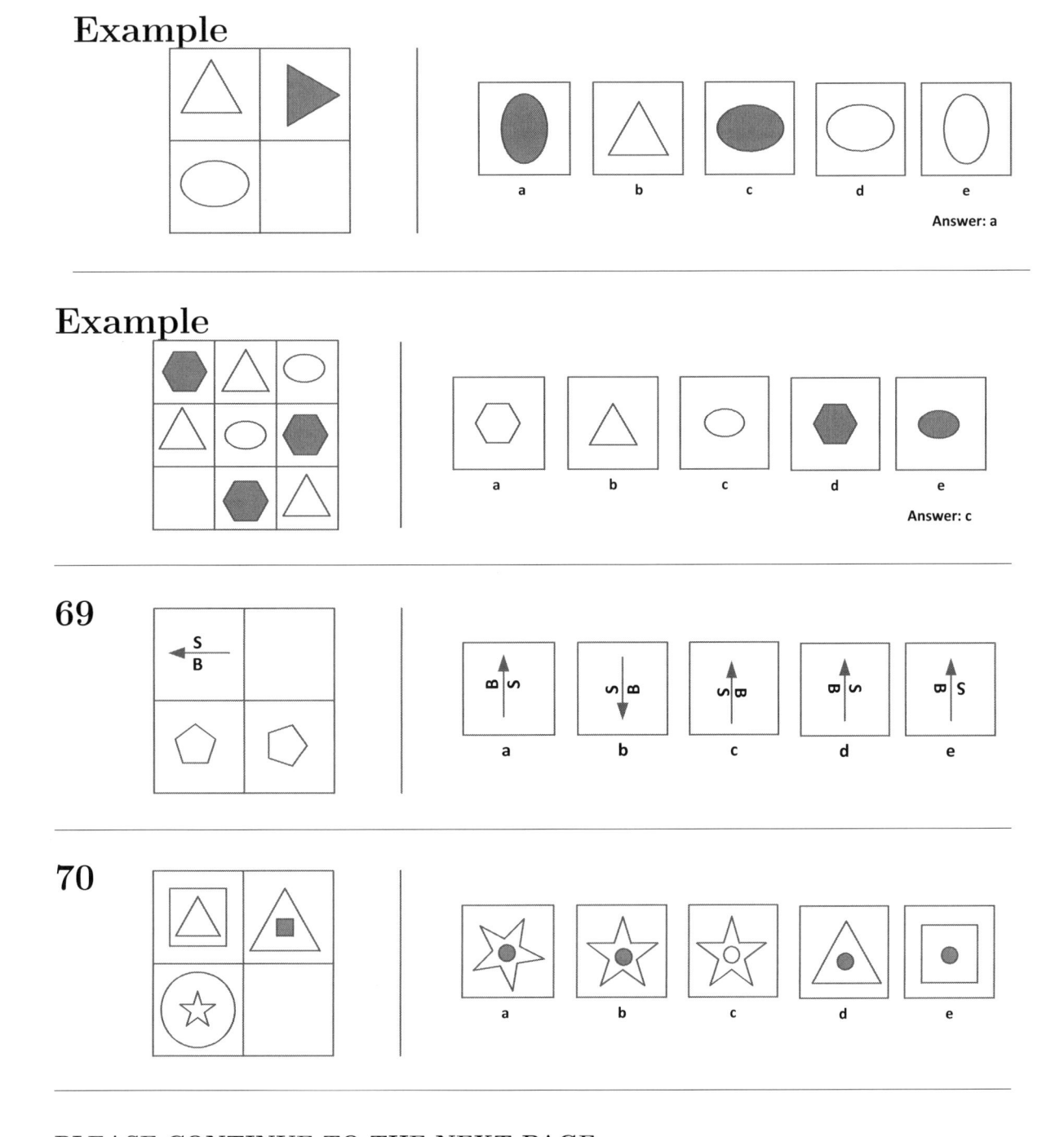

Example

Answer: a

Example

Answer: c

69

70

PLEASE CONTINUE TO THE NEXT PAGE

Mixed Multiple-Choice Pack Three 11B

71.

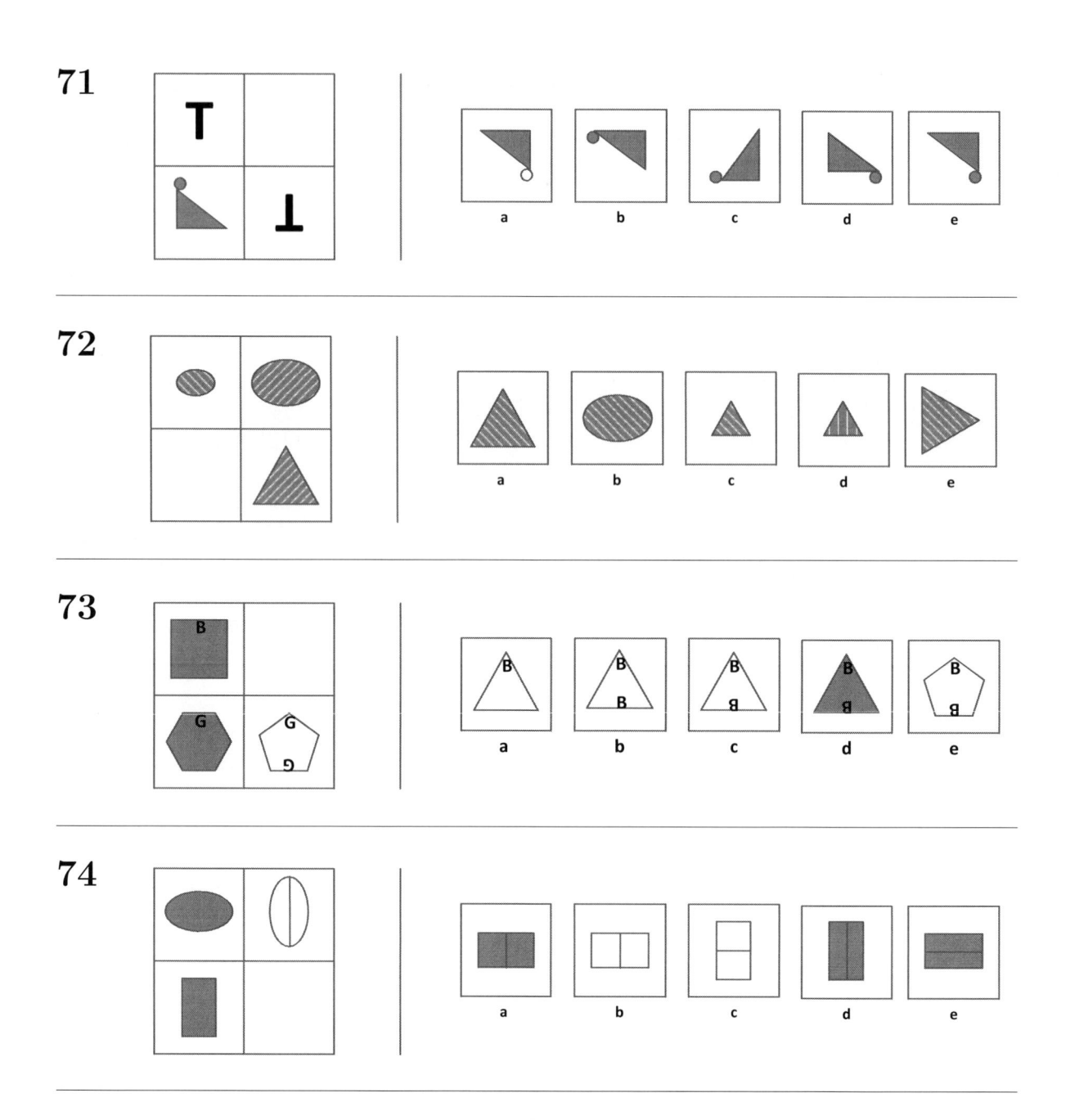

72.

73.

74.

PLEASE CONTINUE TO THE NEXT PAGE

Mixed Multiple-Choice Pack Three 11B

75

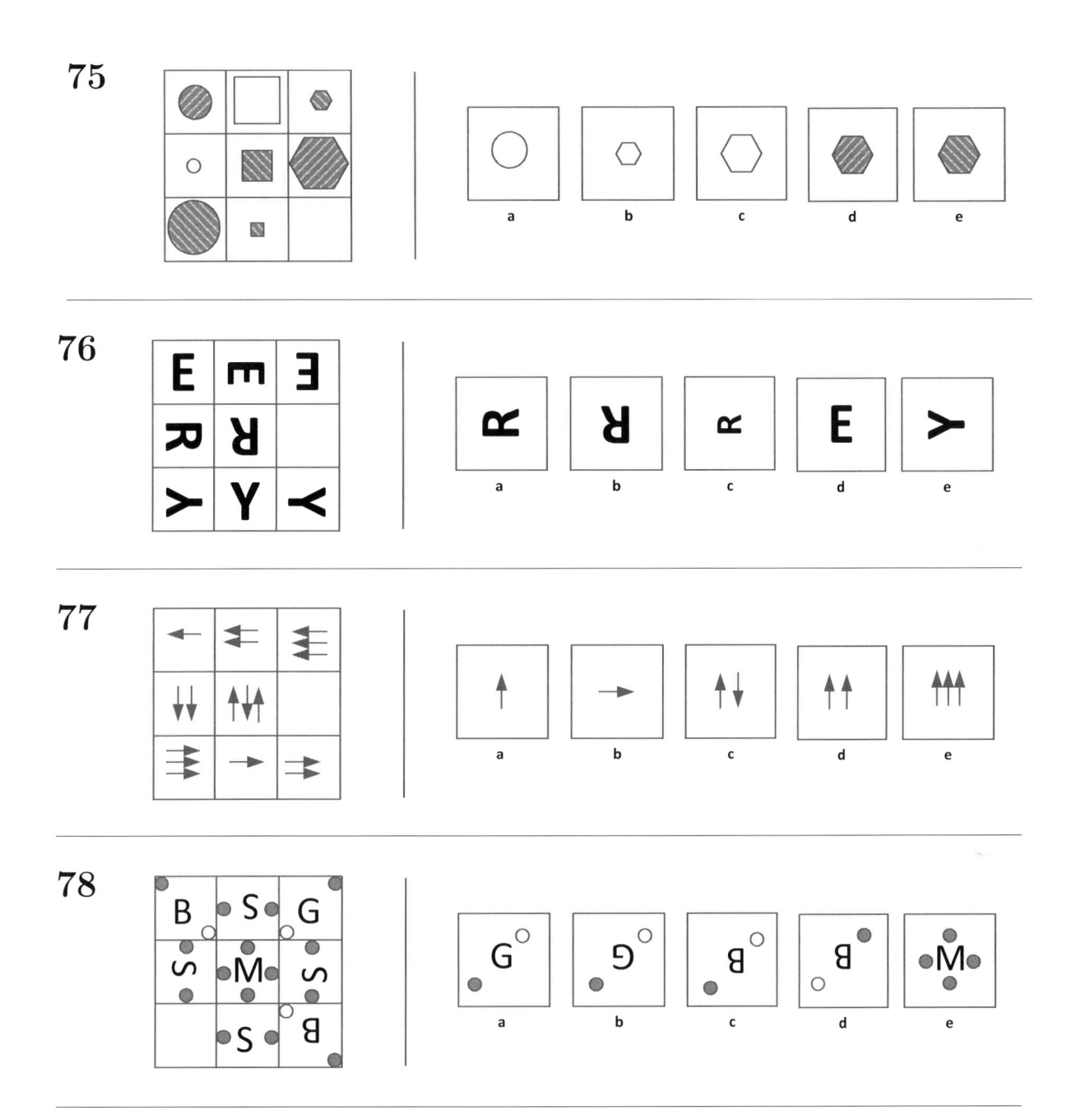

76

77

78

PLEASE CONTINUE TO THE NEXT PAGE

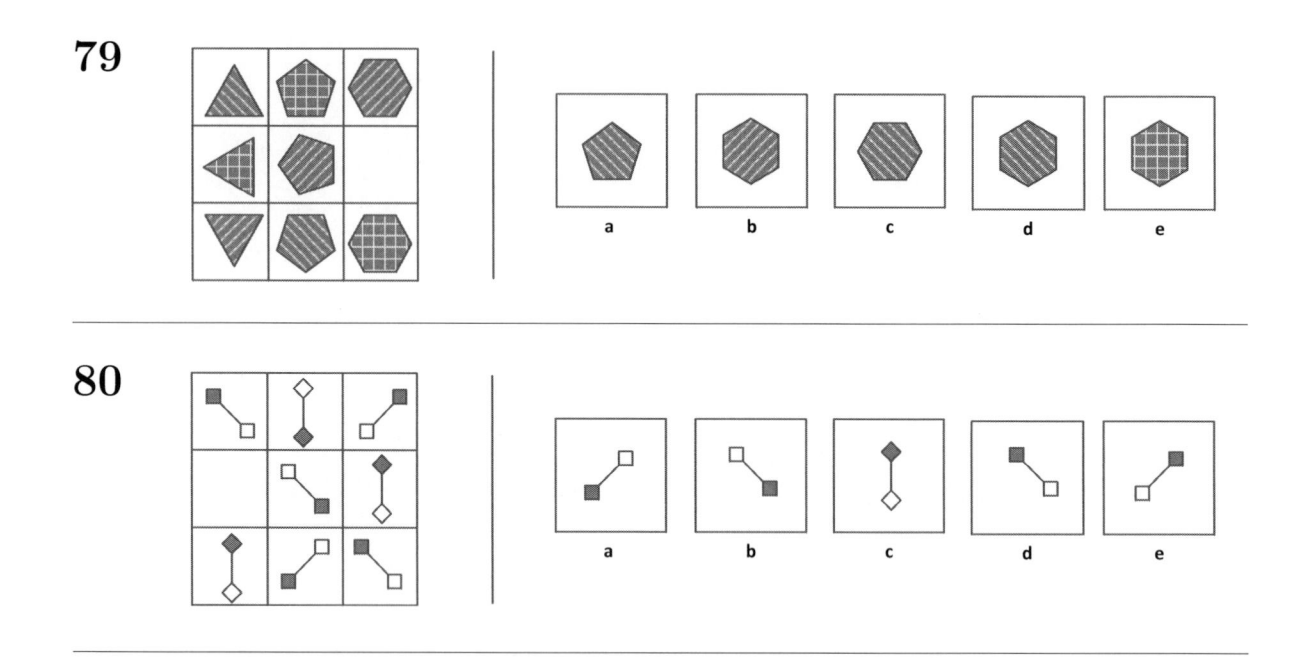

Mixed Multiple-Choice Pack Three 11B © Moon Tuition
www.moontuition.co.uk

79

80

THE END OF NON VERBAL REASONING PART

THE END OF TEST PAPER 11B

11+ Answer Sheets

Multiple-Choice Practice Papers
Pack Three

The following answer sheets are included:

- Multiple-Choice Practice Paper Pack Three 11A.
- Multiple-Choice Practice Paper Pack Three 11B.

Moon Tuition
making the most of your potential

www.moontuition.co.uk

All rights reserved, including translation. No part of this publication may be reproduced or transmitted in any form or by any means, electronic or mechanical, including photocopying, recording, or duplication in any information storage and retrieval system, without permission in writing from the publishers, and may not be photocopied or otherwise reproduced within the terms of any licence granted by the Copyright Licensing Agency Ltd.

Moon Tuition
making the most of your potential

Pack Three 11A - Answer Sheet Page 1

Pupil's Name

School's Name

Date of Test

PUPIL NUMBER

SCHOOL NUMBER

DATE OF BIRTH

Day	Month	Year
(0) (0)	January	2000
(1) (1)	February	2001
(2) (2)	March	2002
(3) (3)	April	2003
(4)	May	2004
(5)	June	2005
(6)	July	2006
(7)	August	2007
(8)	September	2008
(9)	October	2009
	November	2010
	December	2011

Please mark like this ⟵

Questions 1–50: each with options A, B, C, D, E

Moon Tuition
making the most of your potential

Pack Three 11A - Answer Sheet Page 2

Pupil's Name

School's Name

Date of Test

DATE OF BIRTH

Day	Month	Year
(0) (0)	January ☐	2000 ☐
(1) (1)	February ☐	2001 ☐
(2) (2)	March ☐	2002 ☐
(3) (3)	April ☐	2003 ☐
(4)	May ☐	2004 ☐
(5)	June ☐	2005 ☐
(6)	July ☐	2006 ☐
(7)	August ☐	2007 ☐
(8)	September ☐	2008 ☐
(9)	October ☐	2009 ☐
	November ☐	2010 ☐
	December ☐	2011 ☐

PUPIL NUMBER / **SCHOOL NUMBER** (digits 0–9)

Please mark like this ⟵

Moon Tuition
making the most of your potential

Pack Three 11B - Answer Sheet Page 1

Pupil's Name

School's Name

Date of Test

PUPIL NUMBER
SCHOOL NUMBER

DATE OF BIRTH

Day	Month	Year
(0) (0)	January	2000
(1) (1)	February	2001
(2) (2)	March	2002
(3) (3)	April	2003
(4)	May	2004
(5)	June	2005
(6)	July	2006
(7)	August	2007
(8)	September	2008
(9)	October	2009
	November	2010
	December	2011

Please mark like this ←

Questions 1–50, each with options A, B, C, D, E.

Moon Tuition
making the most of your potential

Pack Three 11B - Answer Sheet Page 2

Pupil's Name

School's Name

Date of Test

DATE OF BIRTH

Day	Month	Year
(0) (0)	January	2000
(1) (1)	February	2001
(2) (2)	March	2002
(3) (3)	April	2003
(4)	May	2004
(5)	June	2005
(6)	July	2006
(7)	August	2007
(8)	September	2008
(9)	October	2009
	November	2010
	December	2011

PUPIL NUMBER — digits 0–9

SCHOOL NUMBER — digits 0–9

Please mark like this ←

Questions 51–80, each with options A, B, C, D, E.

11+ Answer Key

Multiple-Choice Practice Papers
Pack Three

Read these instructions before you start marking:

- Only the answers given are allowed.
- One mark should be given for each correct answer.
- Do not deduct marks for the wrong answers.

Moon Tuition
making the most of your potential

www.moontuition.co.uk

All rights reserved, including translation. No part of this publication may be reproduced or transmitted in any form or by any means, electronic or mechanical, including photocopying, recording, or duplication in any information storage and retrieval system, without permission in writing from the publishers, and may not be photocopied or otherwise reproduced within the terms of any licence granted by the Copyright Licensing Agency Ltd.

Mixed Multiple-Choice Pack Three answers

Practice Paper Answers

Practice Paper 11A		Practice Paper 11A	
1. C	26. B	51. C	76. E
2. B	27. A	52. E	77. B
3. D	28. E	53. A	78. C
4. D	29. C	54. B	79. D
5. B	30. B	55. D	80. A
6. A	31. B	56. A	
7. E	32. C	57. C	
8. C	33. A	58. B	
9. B	34. A	59. D	
10. A	35. C	60. E	
11. D	36. B	61. A	
12. C	37. D	62. D	
13. A	38. D	63. B	
14. C	39. A	64. C	
15. E	40. C	65. C	
16. B	41. B	66. E	
17. D	42. C	67. A	
18. A	43. D	68. D	
19. B	44. A	69. E	
20. E	45. D	70. A	
21. B	46. B	71. C	
22. C	47. A	72. D	
23. C	48. C	73. C	
24. E	49. D	74. B	
25. D	50. C	75. B	

Mixed Multiple-Choice Pack Three answers

Practice Paper Answers

Practice Paper 11B		Practice Paper 11B	
1. D	26. A	51. B	76. A
2. C	27. B	52. C	77. A
3. E	28. E	53. A	78. B
4. E	29. E	54. D	79. D
5. B	30. C	55. C	80. E
6. A	31. D	56. D	
7. B	32. B	57. C	
8. C	33. A	58. D	
9. A	34. C	59. C	
10. D	35. E	60. E	
11. B	36. D	61. A	
12. E	37. E	62. D	
13. B	38. A	63. E	
14. D	39. B	64. C	
15. C	40. A	65. C	
16. A	41. D	66. C	
17. C	42. D	67. B	
18. B	43. A	68. A	
19. C	44. B	69. D	
20. E	45. B	70. B	
21. D	46. C	71. E	
22. E	47. A	72. C	
23. A	48. B	73. C	
24. C	49. B	74. B	
25. C	50. A	75. C	

Printed in Great Britain
by Amazon